Osama Awwad

WDM Optical Network Design

Osama Awwad

WDM Optical Network Design

Using Limited Traffic Grooming Resources

VDM Verlag Dr. Müller

Copyright © 2007 VDM Verlag Dr. Müller e. K. and licensors
All rights reserved. Saarbrücken 2007
Contact: info@vdm-verlag.de
Cover image: www.purestockx.com
Publisher: VDM Verlag Dr. Müller e. K., Dudweiler Landstr. 125 a, 66123 Saarbrücken, Germany
Produced by: Lightning Source Inc., La Vergne, Tennessee/USA
          Lightning Source UK Ltd., Milton Keynes, UK

Copyright © 2007 VDM Verlag Dr. Müller e. K. und Lizenzgeber
Alle Rechte vorbehalten. Saarbrücken 2007
Kontakt: info@vdm-verlag.de
Coverbild: www.purestockx.com
Verlag: VDM Verlag Dr. Müller e. K., Dudweiler Landstr. 125 a, 66123 Saarbrücken, Deutschland
Herstellung: Lightning Source Inc., La Vergne, Tennessee/USA
          Lightning Source UK Ltd., Milton Keynes, UK

ISBN: 978-3-8364-3436-2

ACKNOWLEDGMENTS

First and foremost, I thank God for giving me both the opportunity and ability to do this work.

I would specially like to thank my advisor, Professor Ala Al-Fuqaha for his guidance and insight. It has been a pleasure to work with him as a mentor throughout my graduate studies at Western Michigan University. What I value most about the past two years was the opportunity to absorb not only his insights into the specifics of my work, but also a fundamental approach to research that emphasizes rigor and first principles. I hope this thesis will go some way towards expressing my thanks.

I would like to thank my thesis committee members: Professor Mohsen Guizani, and Professor Dionysios Kountanis. I had the pleasure of learning from them in many ways as well. The path to my Master's degree would have been a lot less rewarding without Dr. Guizani. He has taught me a great deal over the last two years. While, Dr. Kountanis has been a friend, an insightful in this exciting field.

I am also indebted to Professor Donna Kaminski for her tremendous ongoing help and support, and for the fun and motivating discussions.

Ghassen Ben Brahim, Wasim El-Hajj, and Adawia Al-Alawnih are fellow students at Western Michigan University, it was a pleasure for me to work with them. I especially would like to thank Ghassen Ben Brahim for the work and discussions we made on this project.

Acknowledgements—Continued

Also, I would like to extend my appreciation to Dr. Abdella Battou (VP of Engineering and CTO of Lambda Optical Systems) for giving me the opportunity to gain hands on experience in the optical networking industry.

I would like to thank Ahmad Shaban, Arqam Al-Shaikhly, Ashraf Ghanam, Enad Mahmoud, Fouad Alaqad, Ibrahim Abd rabou, Hussam Khasawneh, Maher Al-Tayyeb, Mohd Khater, Mohd Alajmi, Yazan Al-Wedyan, Mutaz Daana, Mohd Walid, Mohd Jaber, and Mustafa Al-Adarbeh for being the most faithful friends. Also, I would like to thank Dr. Abubaker and his family for their kind hospitality during my stay in Kalamazoo.

Last but not least, I would like to extend my whole-hearted thanks to my family. Special thanks to my parents, for their love and support throughout my studies. I especially would like to thank my brothers Yousef and Ayman for their guidance. Finally, I would like to dedicate this work to my brother Mohammed for listening to me, encouraging me, and putting me on the right track.

Osama Awwad

TABLE OF CONTENTS

Table of Contents—Continued

Table of Contents—Continued

CHAPTER

LIST OF TABLES

LIST OF FIGURES

List of Figures—Continued

CHAPTER 1

INTRODUCTION

Over the past few years, the field of computer and telecommunication networks has experienced tremendous growth. Traffic demand has increased substantially, somewhat unexpectedly, prompting carriers to add capacity quickly and in the most cost effective way possible. This change in the fundamental character of backbone network traffic prompted Internet Service Providers (ISPs) and Internet Backbone Providers (IBPs) to switch to optical transmission technology by replacing the traditional capacity limited copper cables with optical fibers. This radical change required also a modification in all underlying communication protocols.

High data rate, noise rejection, and electrical isolation are some of the main features that made optical transmission the technology of choice for all major telecommunication carriers. Nowadays, most of the optical transmission equipments is still based on the electronic processing of optical signals, which requires the optical signal to go through several processing stages before it reaches its intended destination. Using this equipment, the optical signal is converted to an electrical signal, amplified, switched, and finally the electrical signal is reconverting back to optical domain. This is generally referred to as Optical-Electrical-Optical (O-E-O) conversion. O-E-O equipment presents a significant bottleneck in today's transport networks. Therefore, it is in the interest of ISPs and IBPs to replace existing O-E-O equipment with all optical one [4], and avoid going through costly O-E-O processing

1

stages. Moreover, O-E-O transmission equipment puts a bound on the signal processing power because of their limited ability to process the electrical signal in acceptable time. For example, an O-E-O amplifier that was state of the art several years ago may not be able to keep up with the demands of the future. However, an **all-optical** amplifier does not set any bounds or restrictions on the signals that need to be amplified.

All-optical equipment (sometimes referred to as Optical-Optical-Optical or O-O-O) switch the optical signal to a different output without the need for O-E-O (Optical-Electrical-Optical) conversion. All optical switching equipment can be implemented using different technologies. These technologies include liquid crystals, holographic crystals, tiny mirrors, etc. One of the most widely used technologies by all-optical equipment manufacturers is the tiny moveable mirrors known as Micro-Electro-Mechanical Systems (MEMS). MEMS consist of mirrors no larger in diameter than a human hair arranged on special pivots giving them the freedom to move in three dimensions. Thanks to the advances in this kind of technology, mirror arrays of no larger than a few centimeters square can support hundreds of mirrors. Light from an input fiber is aimed at a mirror, which is directed to reflect the light to another mirror on a facing array. This mirror then reflects the light down towards the desired switch output. [14]

One of the major advantages of building all-optical networks is network scalability. All-optical equipment provides traffic multiplexing capability that is bit-rate and protocol-independent. New capacity can be added to the O-O-O network simply by adding a new fiber link without replacing the entire network infrastructure. The scalability of all-optical networks, the physical security, and the high data rate features have made from the all optical technology a potential basis for future network infrastructure.

Optical fibers can carry multiple data streams by assigning each to a different wavelength. This approach is known as Wavelength Division Multiplexing (WDM). Currently, WDM is classified as: (1) coarse WDM (CWDM) with ≈ 40 wavelengths per fiber and (2) dense WDM (DWDM) with ≈ 200 wavelengths per fiber. Each wavelength can be viewed as a channel that provides an optical connection between two nodes. Such a channel is called a lightpath or a connection. A lightpath may span multiple fiber links, e.g., provide a "circuit-switched" interconnection to support a heavy traffic flow between two nodes located far from each other in the physical transmission network. Each intermediate node on the light path essentially provides an all-optical bypass facility to support the lightpath. Once a set of lightpaths has been determined, each lightpath needs to be routed and assigned a wavelength This is referred to as a routing and wavelength assignment (RWA) problem. [1].

Based on the type of network traffic, the RWA problem can be classified into two categories: RWA with static network traffic and RWA with dynamic network traffic. In the static case, the set of connections between the source and destination pairs is known in advance and a lightpath needs to be established for each connection (i.e., offline RWA). However, in the dynamic case, routing and wavelength assignments are done on the fly as lightpath requests arrive to the network (i.e., online

3

RWA). In this case, routing and wavelength assignment decisions are based on the current network state.

The RWA problem was proven to be insufficient to ensure the most efficient utilization of network resources [5,6]. In order to overcome the aforementioned deficiency, researchers are evaluating the cost and performance of multiplexing low speed traffic streams into high capacity ones before assigning then wavelength resources. This technique is referred to the RWA problem with traffic grooming (GRWA.) Figure 1.1 is an illustration of the GRWA problem. In this figure, we show one OC-1 (51 Mbps) and one OC-12 optical signals that are being multiplexed into a higher-rate OC-48 carrier. This is realized by time division multiplexing (TDM) technology, which uses different time slots on a high-rate channel to transmit different lower-rate data signals.

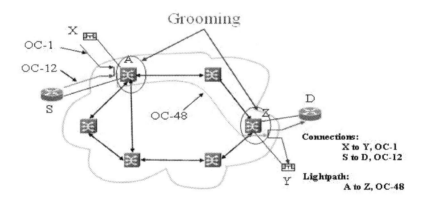

Figure 1.1: Traffic grooming illustrated

Most previous research on traffic grooming in WDM mesh networks assumes that traffic grooming capabilities are possible throughout the optical network domain or at the edge nodes only. Unfortunately, these approaches may not be practical or cost-effective. In this work, due to the high cost of traffic grooming devices, we allow a few nodes to support traffic grooming.

In addition, because of the high cost of all-optical wavelength conversion resources, we collocate the wavelength conversion and traffic grooming resources on the same node since optical transponders are capable of traffic grooming and wavelength conversion at the same time. This allows us to eliminate the wavelength continuity constrain at transponder equipped nodes and thus, significantly improve the network blocking performance. We call this an optical network with sparse traffic grooming and wavelength conversion resources.

Previous research work on traffic grooming focused on (1) maximizing network utilization, (2) maximizing the traffic demand that can be carried given the network physical constraints, and (3) minimizing the network cost. In order to achieve the objectives mentioned above, researchers put many restrictions on the network topologies and allowed traffic grooming to exist only at the end nodes; this type of grooming is known as single-hop traffic grooming [5]. If we remove this restriction, then we can have grooming at any node throughout the optical network; we denote the general grooming case as multihop traffic grooming [5].

In this work, we focus on the traffic grooming problem in optical networks with sparse traffic grooming and wavelength conversion capabilities. We study multihop traffic grooming and allow nodes to have no traffic grooming support.

Furthermore, we assume that traffic grooming and wavelength conversion resources are collocated and limited in number.

We wish to achieve several (possibly mutually exclusive) network properties with our model. In particular, we seek to minimize the cost of traffic grooming and conversion hardware, while minimizing the blocking probability. Furthermore, propose GRWA heuristics that strive to maximize the utilization of the network while minimizing the number of wavelengths needed.

The rest of this thesis is organized as follow: Chapter 2 provides a literature survey about optical networks and traffic grooming. Chapter 3 formally introduces the traffic-grooming problem (GRWA) and presents an Integer Linear Programming (ILP) formulation of the problem. Chapter 4 presents our most-contiguous heuristic to solve the GRWA problem in networks with sparse resources under static and dynamic traffic patterns. Chapter 5 introduces a genetic approach to solve the GRWA problem in optical networks with collocated traffic grooming and wavelength conversion resources. Chapter 6 presents our numerical and simulation results. Finally, Chapter 7 concludes our study and discusses possible future research extensions.

CHAPTER 2

WDM OPTICAL NETWORK CONTEXT

2.1 Introduction

Telecommunication networks in general, can be divided into three major

parts: the access network, the metropolitan-area network, and the Long-Haul transport

Network. These are better illustrated in Figure 2.1.

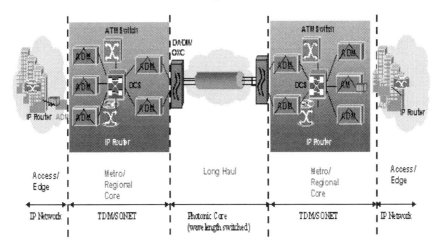

Figure 2.1: Telecommunication network structure

Access networks are that portion of a public switched network that connect

access nodes to individual subscribers. More simply, they are the last link in a network between the customer premises and the first point of connection to the network infrastructure—a point of presence (PoP) or central office (CO). Widely used technologies for access networks include dial-up modems, Asymmetric digital subscriber lines (ADSL), cable modems.

Metropolitan Area Networks (MANs) are the second level of the Internet hierarchy, connecting access networks to the backbone. MANs typically cover an area between 10 to 100 Kilometers in diameter. It mainly employs Synchronous Optical Network (SONET) in the U.S. or Synchronous Digital Hierarchy (SDH) in Europe using point-to-point or ring topologies with add/drop multiplexers (ADMs).

Long-haul networks (sometimes referred to backbone networks) are the top level in the Internet hierarchy. Their function is to connect different MANs using high-speed data transmission, so the primary concern in such networks is efficient capacity utilization. Due to the inefficiency and poor scalability of interconnected rings, backbone networks are expected to migrate to resource-efficient and scalable meshes.

The current trend in developing networking systems for the network core is based on either optical packet switching or optical wavelength switching. While optical packet-switched networks are somewhat futuristic, wavelength-switches optical networks are becoming realistic to deploy.

The rest of this chapter is organized as follows. The next section, provides background information about WDM network. Next, we present the traffic grooming problem in WDM networks. Finally, we introduce the genetic algorithm as a general heuristic to solve optimization problems encountered in WDM optical networks.

2.2 WDM Network Background

2.2.1 WDM Architecture

The architecture of wide-area WDM networks that is expected to be the basis for future all-optical infrastructure is based on the concept of *wavelength routing*. A wavelength-routed optical network consists of photonic switching fabrics connected by a set of fiber links to form an arbitrary physical topology. In such networks, each end-user is connected to a switch via a fiber link. The combination of an end-user and its corresponding switch is referred to as a network node. Each node is equipped with a set of transmitters and receivers, which may or may not be wavelength tunable [1].

In a wavelength-routed network, lightpath requests define a logical topology. A lightpath is defined as a clear all-optical channel between two nodes that may traverse more than one fiber link in the optical network.

In WDM network, nodes are equipped with optical cross-connects (OXC) devices that switch wavelengths from the switch input to the output ports, enabling the establishment of direct lightpath connections between any pair of nodes. Upon the arrival of optical signals with different wavelengths at different input ports, the OXC device, independently switches each signal to the appropriate OXC output port. An OXC with N input and N output ports capable of handling W wavelengths per port can be thought of as W independent $N \times N$ optical switches.

Currently, telecommunication carries can deploy two different types of OXC switches: converter capable and non-converter capable switches. In the absence of any wavelength converter device, each OXC uses the same wavelength as that of the incoming signal when switching the optical signal from the input to the output port. Such constraint is referred to as the wavelength continuity property. In the presence of wavelength converter devices, this constraint is no longer applicable, i.e. the incoming and outgoing optical signals may have different wavelengths. A wavelength converter is a single input/output device that converts the wavelength of an optical signal arriving at its input port to a different wavelength as the signal departs from its output port. Figure 2.2 illustrates the difference between converter capable and non-converter capable OXCs.

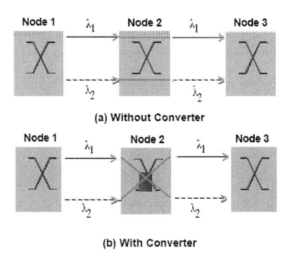

Figure 2.2: Wavelength conversion

2.2.2 WDM Network Control Plane

Currently, one of the most pressing issues in WDM optical networks is how to manage and control such large networks. Conceptually, the optical network has three major control planes as shown in Figure 2.3:

1. Transport Plane: provides high-speed data transmission.

2. Control Plane: provides real-time signaling and routing.

3. Management Plane: manages the network resources, and monitors network state.

Inside an administrative domain, OXCs are interconnected through fibers in a mesh topology, and are able to communicate with one another through the Internal Network-Network-Interface (I-NNI). The communication between different administrative domains is through the External Network-Network-Interface (E-NNI).

Upon the initiation of a lightpath request by the network management system or by a client through the User Network Interface (UNI), a route computation process starts [15]. In case an eligible route is found, the control plane signals the control unit on the OXC and sets up the lightpath by communicating with the other OXCs. This lightpath establishment is typically implemented over administratively configured ports at each OXC and uses a separate control wavelength on each fiber. Thus, we distinguish between the paths that data and control signals take in the optical network: data lightpaths originate and terminate at client subnetworks and, transparently traverse the OXCs, while control lightpaths are terminated at the control unit of each OXC [31].

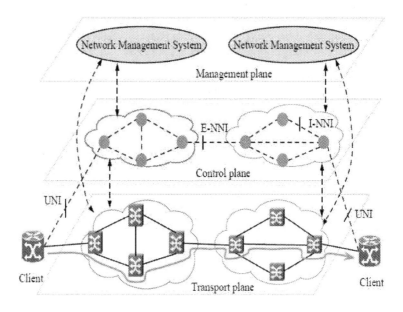

Figure 2.3: Optical network overlays

2.3 Traffic Grooming in WDM Network

2.3.1 Traffic Grooming Background

While a single fiber strand has over a terabit-per-second bandwidth and a wavelength channel has over a gigabit-per-second transmission speed, the network may still be required to support traffic connections at rates that are lower than the full wavelength capacity. The capacity requirement of these low-rate traffic connections can vary in ranges from STS-1 (51.84 Mbps or lower) up to full wavelength capacity. In order to reduce deployment costs and improve network performance, it is important

12

for network operators to be able to *"groom"* the multiple low-speed traffic connections into high-capacity circuit pipes.

In wavelength-routed optical networks without traffic grooming devices, lightpaths are established by assigning distinct wavelengths. This wavelength assignment constraint requires each connection be carried over a distinct wavelength. However, when nodes have traffic grooming capabilities, multiple lightpath requests can be multiplexed together and assigned a single wavelength. Figure 2.4 (b) illustrates the traffic grooming process during lightpath establishment.

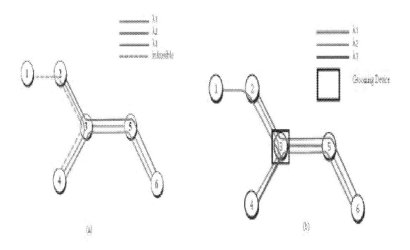

Figure 2.4: Traffic grooming concept illustrations

In the network model shown in Figure 2.4, each span has an identical capacity of 3 wavelengths and there is a demand of 1 wavelength between the following

13

source-destination pairs: (2,4), (3,5), (5,6), (2,6), (6,4) and (1,4). The routing for each

channel is unique and there is sufficient wavelength for each request except (1,4),

which cannot be accommodated as illustrated in Figure 2.4 (a). Hence, the routing and

wavelength assignment problem is infeasible. The solution for this problem is to use a

traffic grooming device at node 3. Figure 2.4 (b) illustrates that a traffic grooming

device installed on node 3 can multiplex the traffic carrier on lightpath (1,4) with that

carrier on lightpath (6,4). In this example, we assume that $\lambda1$ capacity is large enough

to carry the aggregated requests.

In the next section, we review recent work on traffic grooming in optical

networks with more emphasis on mesh optical networks.

2.3.2   Traffic Grooming Literature Survey

a. Traffic Grooming in SONET Ring Network

Much of today's physical layer transmission infrastructure is built around

Synchronous Optical Network (SONET) rings. In a SONET ring network, WDM is

mainly used as a point-to-point transmission technology. SONET multiplexers have

the ability to "groom" lower rate SONET signals into a single high rate SONET

stream. For instance, four OC-3 circuits can be multiplexed together into an OC-12

circuit and 16 OC-3's can be multiplexed into an OC-48. Electronic add-drop

multiplexers (ADMs) are used to add/drop traffic at intermediate nodes to/from the

high-speed channels.

Due to the wide deployment of SONET/SDH technology, traffic grooming in

14

networks with ring topology, has been widely discussed in several research papers [6, 7, 8, 9, 10]. The major cost of such a network is considered to be dominated by SONET ADMs. Therefore, most of the recent research work has focused on minimizing the number of SONET ADMs that need to be deployed.

The traffic grooming problem has been proven to be NP-complete [16] in ring optical networks even in the presence of full wavelength conversion capabilities.

As a network design problem, the authors in [17] consider a special case of the traffic grooming problem in unidirectional SONET/SDH ring networks and attempt to minimize the network cost dominated by SONET ADMs. Heuristic algorithms to achieve this objective were presented for special traffic patterns such as uniform, and certain cases of cross-traffic. Moreover, the authors consider the use of hub nodes, where traffic can be switched between SONET rings and show that, networks using hub nodes require the same number of ADMs compared to networks that do not use hub nodes.

b. Traffic Grooming in WDM mesh network

Upon the migration of optical backbone networks from ring to mesh topology and the considerable growth in Internet traffic, traffic grooming in WDM mesh networks becomes an extremely important area of research. Moreover, mesh networks are more flexible to various network failures and more flexible in accommodating changes in traffic demands [21, 22]. In [19], the authors address the network design problem in both mesh and ring networks. An Integer Linear Program (ILP) formulation and two heuristic algorithms are proposed for mesh and ring network designs. In [19], the authors conclude that mesh networks are more resilient to various network failures than ring topology networks and have cost advantage for sufficiently large distance scale networks.

In [5], the authors propose several node architectures for supporting traffic grooming in WDM mesh networks. They formulate the static traffic grooming problem for single-hop and multi-hop networks as an ILP problem and present two heuristic algorithms to compare the performance with that of the ILP.

In [12], the authors consider the traffic grooming problem with the objective of minimizing the number of transponders in WDM mesh network. The problem is first formulated as an ILP problem. Because it is very hard to find a solution for large networks, the authors reduce the size of the ILP problem by proposing a decomposition method that divides the traffic grooming problem (GRWA) into two smaller problems: the traffic grooming and routing problem (GR), and the wavelength assignment problem (WA). The GR problem is formulated as an ILP problem, while heuristic algorithms are proposed to solve the WA problem. Despite of using the decomposition technique, the ILP formulation still cannot be directly applied to large networks. Moreover, this approach requires all traffic requests to be known in advance, which cannot be satisfied in dynamic grooming.

Contrarily to the aforementioned research work, where the authors consider only static traffic, the authors in [18] addressed the issue of using fixed-alternate routing during the dynamic traffic grooming. The objective is to satisfy as many connections as possible in the network, leading to a high network throughput and low network blocking probability. An online algorithm, namely, fixed-order grooming (FOG) is proposed. The FOG algorithm can be used for both single-hop traffic grooming and multi-hop traffic grooming.

As WDM optical networks migrate from ring to mesh topologies, it is important to solve the traffic-grooming problem in networks with sparse resources. In a sparse traffic grooming network, some nodes may have traffic grooming capabilities while others may not have any (traffic must stay on a distinct wavelength when flowing through these nodes). This problem was addressed in [13], where the authors presented an ILP formulation and a heuristic approach to solve the grooming node placement problem in sparse grooming networks under static traffic. Contrarily to our study, this work does not support networks with sparse wavelength conversion resources. It is assumed that all the nodes in the optical network either have grooming capabilities or not, while in our work, we impose constraints on the grooming capabilities in terms of the number of transceivers used for originating and terminating optical lightpaths.

The dynamic traffic grooming with sparse capabilities was also studied in [20]. In that work, the authors propose two algorithms to exploit efficiently the sparse grooming capability that exist in the network under multi-granularity traffic. The authors conclude that optical networks with sparse traffic grooming resources provide an effective and economical solution for telecommunication carriers

2.4 Genetic Algorithm Background

Genetic algorithms are a particular class of evolutionary algorithms that use techniques inspired by evolutionary biology such as inheritance, mutation, natural selection, and recombination (or crossover). Genetic algorithms are typically implemented as a computer simulation to solve optimization problems. Genetic Algorithms (Gas) start with a population of abstract representations (called *chromosomes*) of candidate solutions (called *individuals*) and evolves toward better solutions. Traditionally, solutions are represented as binary strings of 0s and 1s, but different encodings are also possible. The evolution starts from a population of completely random individuals and happens in generations. In each generation (i.e., epoch), the fitness of the whole population is evaluated, multiple individuals are stochastically selected from the current population (based on their fitness), modified (mutated or recombined) to form a new population, which becomes the current generation in the next iteration of the algorithm. [28]

2.4.1   Genetic Algorithm Structure

a. Encoding

Encoding of chromosomes is the first question to ask when starting to solve a problem with GA. There are different ways of encoding. The encoding depends mainly on the problem under study.

b. Initial Population

A genetic algorithm starts with an initial population of strings that will be used to generate successive populations of strings afterwards. The initialization is usually done randomly or by a heuristic algorithm.

c. Reproduction

The reproduction creates a new population by repeating the following steps over and over to generate successive generations of more "fit" solutions until a handful of feasible solutions remain:

1. *Evaluation*: After every generated population, every individual of the population must be evaluated with the goal of distinguishing between good and bad individuals. This is done by mapping the objective function to a 'fitness function'

2. *Selection*: chromosomes are selected from the population to be parents for crossover. The philosophy behind the selection of the chromosomes is based on Darwin's theory of evolution, which favors the best ones to survive in order to create new offspring. There are many methods in selecting the best chromosomes, such as: the roulette wheel selection, the tournament selection, the rank selection, the steady state selection, etc.

3. *Crossover*: Once two parents were selected, the genetic algorithm combines them to create two new offspring. The combination is performed by the crossover operator, which allows the advantageous

19

traits to be spread throughout the population with the goal of having the whole population benefit from this chance discovery.

4. *Mutation*: After a crossover is performed, mutation takes place in order to truly emulate the genetic process. A mutation operator needs to be incorporated in order to account for the random mistakes that may occur. This is done by occasionally flipping values, which introduces new features into the population pool.

5. *Accepting*: Place new offspring in the new population based on the fitness function.

2.4.2   Genetic Algorithm Related Work

GA has been proven to be a practical and robust optimization and search tool for network design [23, 24], therefore it also a promising approach to solve Routing and wavelength assignment problems in WDM networks.

In [25], the authors formulate the RWA problem as an optimization problem and solve it using genetic algorithms. In their approach, each gene in an individual represents one of the *k*-shortest paths between the source and destination nodes. This approach solves the routing problem; however the wavelength assignment is done using a heuristic algorithm.

In [26], the authors employ a genetic algorithm for traffic grooming in WDM networks by optimizing a single objective function. The objective is to assign wavelengths to incoming traffic connections such that the overall network cost is

20

minimized. The overall cost includes the cost of transceivers at the nodes and the number of wavelengths. This work considers only the wavelength assignment problem for traffic grooming, but it doesn't solve the routing problem.

In [27], a model based on a combination of genetic algorithms and clustering heuristics is employed to solve the traffic grooming problem in WDM mesh network. The routing problem is solved using the GA model while traffic grooming is solved based on the clustering heuristic. The objective of that research work was to maximize the lightpath utilization and to minimize the network cost. An encoding scheme called Position based Bit Representation (PBR) was used. In this encoding scheme, each gene of a chromosome is coded as a single bit where each bit represents an edge. Thus, the PBR representation for the routing path is constructed from the genes with value 1 in the chromosome.

CHAPTER 3

ILP FORMULATION OF THE GRWA PROBLEM IN WDM NETWORKS
WITH SPARSE RESOURCES

3.1 Introduction

In this chapter, we formulate the static GRWA problem in optical networks with sparse traffic grooming and wavelength conversion resources as an integer linear programming (ILP) problem. Our formulation considers two possible objective functions: (1) Minimize the total number of hops used by all incoming lightpath requests, and (2) Minimize the total cost of traffic grooming and wavelength conversion equipment. The chapter also presents numerical results obtained from a program that we implemented using ILPSolve (an implementation of the SIMPLEX algorithm in JAVA) to verify the correctness of our mathematical formulations under various topologies and traffic scenarios.

3.2 Problem Statement

Our formulation relies on several assumptions:

1. The network topology is a mesh with directed fiber connections. At most two fibers (one in each direction) can connect a pair of nodes.

2. The network switches may have full traffic grooming and wavelength conversion capabilities. However, it is possible to require nodes to have to traffic grooming of wavelength conversion resources.

3. At any given node, we have the required optical receivers and transponders for the used wavelengths (provided the wavelength assignment is valid)

4. Lightpaths do not contain loops. We assume that the routing for a connection can be done using one of the paths given by the K-shortest paths algorithm.

5. The enumeration of all possible lightpaths is done by taking all the routes generated by the K-shortest paths algorithm for each source-destination pair. After all the routes are generated, all possible wavelength assignment combinations are generated. Each unique wavelength assignment on a route is considered as a unique lightpath. We note that lightpaths cannot change wavelengths on the set of nodes that do not have wavelength conversion devices.

Our formulation requires the optical network graph and the lightpath connection requests to be provided as input. The graph of the network is given as a set of edges and vertices $(G = (V, E))$. The requested connections are given by a matrix for each desired connection size, with each element specifying the number of connections (of that size) for that source-destination pair. If desired, one or more vertices may be forced not to have any traffic grooming equipment.

The input is then preprocessed to provide the given network topology and requested lightpath connections to our ILP formulation. The preprocessing involves setting up many matrices including the lightpath-connection and lightpath-link incidence matrices. In order to find the possible lightpaths, an implementation of the K-shortest paths algorithm is used to find the K-shortest routes for a given source-destination pair. Given a route, many lightpaths are generated by considering each possible permutation of wavelength assignment as a unique lightpath. Of course, we do take advantage of the fact that all nodes without traffic grooming resources also have no wavelength conversion resources. After the preprocessing stage, an implementation of our mathematical formulation using ILPSolve is used to obtain numerical results. ILPSolve is an ILP solver engine that provides an implementation of the SIMPLEX algorithm in JAVA.

3.3 Resource Utilization Formulation

Assume:

- lm and mn : Start-end node pairs for a physical fiber link. In addition, we enforce $l \neq m \neq n$ at all times.

- s, d : Source and destination nodes, respectively, of a requested connection.

- i, j : In general the row and column indices of a matrix.

- w : A particular wavelength.

- c : A particular connection size.

- P : Number of all possible lightpaths between source and destination nodes.

Given:

- C_{max} : Capacity of one wavelength on one fiber.

- $C = [1; 3; 12; ...; C_{max}]$: Capacities of connection sizes.

- L : Number of links.

- W : Number of wavelengths per fiber.

- N : Number of nodes.

- N_{sd} : Number of source-destination pairs.

- $D = [d_i]$ Vector of length P , where

$$d_i = \text{number of links used by path } i.$$

- $\phi = [\phi_m]$ Vector of length N , where

$$\phi_m = \begin{cases} 1 \text{ if node } i \text{ has no grooming devices} \\ 0 \text{ otherwise} \end{cases}$$

- $\Lambda = [\lambda_{ij}]$: Requested connections matrix of size N_{sd} x $|C|$, where

$$\lambda_{ij} = \begin{cases} n \text{ if } n \text{ conns. of size } c_j \in C \text{ are req.} \\ 0 \text{ otherwise} \end{cases}$$

- $A = [a_{ij}]$: The P x N_{sd} lightpath-connection incidence matrix, where

$$a_{ij} = \begin{cases} 1 \text{ if lightpath } i \text{ is between } sd \text{ pair } j \\ 0 \text{ if lightpath } i \text{ is not between } sd \text{ pair } j \end{cases}$$

- $G^w = [g_{ij}^w]$: A set of W P x L lightpath-link incidence matrices, where

$$g_{ij}^w = \begin{cases} 1 \text{ if light path } i \text{ uses wavelength } w \text{ on link } j \\ 0 \text{ if light path } i \text{ doesnt use } w \text{ on link } j \end{cases}$$

Variables

- $X = [x_{ij}]$: The path variable matrix with size $P \times |C|$, where

$$x_{ij} = \begin{cases} n \text{ if lightpath } i \text{ has } n \text{ conns. of size } c_j \\ 0 \text{ if lightpath } i \text{ has no conns. of size } c_j \end{cases}$$

- $S = [s_i]$: Vector of length P, where

$$s_i = \sum_j x_{ij}$$

- y_{mn}^{wd}: Indicator variable for route and wavelength assignment of traffic introduced on the nodes. Given a node m and routing wavelength wd we have for each link.

$$y_{mn}^{wd} = \begin{cases} 0 \text{ if no lp starts at } m \text{ and uses } wd \text{ on } mn \\ 1 \text{ if an lp starts at } m \text{ and uses } wd \text{ on } mn \end{cases}$$

- $y_{lmn}^{ws \, wd}$: Indicator variable for route and wavelength assignment of traffic on the nodes. Given an incoming wavelength ws and outgoing wavelength wd, node m, and incoming link lm, we have for each outgoing link mn :

$$y_{mn}^{wd} = \begin{cases} 0 \text{ if no lp uses ws on } lm \text{ and } wd \text{ on } mn \\ 1 \text{ if lp uses ws on } lm \text{ and } wd \text{ on } mn \end{cases}$$

Optimize Minimize the total number of hops used by all the routed connections.

Minimize $S.D$ (1)

26

Subject to

$$A^T X \geq \Lambda \qquad\qquad (2)$$

$$\sum_{1 \leq j \leq |C|} c_j \; col_j(X) .* col_k(G^w) \leq C_{max} \quad \forall k, w \quad (3)$$

$$S . (col_{lm}(G^{ws}) .* col_{mn}(G^{wd})) = \psi_{lmn}^{ws \; wd} \qquad (4)$$

$$y_{lmn}^{wswd} \leq \psi_{lmn}^{ws \; wd} \qquad\qquad (5)$$

$$C_{max} \; y_{lmn}^{wswd} \geq \psi_{lmn}^{ws \; wd} \qquad\qquad (6)$$

$$S . (col_{mn}(G^{wd})) - \sum_{l, ws} \psi_{lmn}^{ws \; wd} = y_{lmn}^{wd} \qquad (7)$$

$$\phi_m \sum_{mn, wd} y_{lmn}^{wswd} \leq 1 \qquad\qquad \forall lm, ws \quad (8)$$

$$\phi_m (y_{lmn}^{wd} + \sum_{lm, ws} y_{lmn}^{wswd}) \leq 1 \qquad\qquad \forall mn, wd \quad (9)$$

Explanation of Equations: We desire to minimize the number of hops used by all the nodes in the network. We start by enumerating all the possible lightpaths, and then, impose our desired conditions on the selected lightpaths. The objective function to minimize is (1). Inequality (2) requires the number of routed connections for a given source destination pair to be greater than or equal to the number of requested connections for that pair. (3) requires the sum of the sizes of the connections on any channel to not exceed the channel capacity. We use (4) to substitute for the expression on the left hand side in the next inequalities. (5) and (6) are used to make the y variables boolean and exist for each fixed set of ws, wd, l, m, n. Inequality (7) gives variables that express how many connections were added at a given node and sent out on a given channel and exists for each fixed set of wd, m, n. Nodes without grooming devices cannot demultiplex connections (8) or multiplex connections (9). Wavelength conversion on nodes without grooming

27

devices is precluded by the enumeration of the lightpaths.

3.4 Network Cost Formulation

In order to formulate a cost based objective function, we assume that the main cost for the traffic grooming enabled switches comes from adding connections, dropping connections, and wavelength conversion. The cost for grooming is α times the number of groomed connections and β times the number of wavelength conversions. The statement of the cost based ILP requires all of the utilization specification presented in the previous section except for D and the optimization function. Here we re-define D and provide a new optimization function.

Optimize Minimize the total cost of the grooming and wavelength conversion equipment. We assume that $\alpha < \beta$ to reflect typical equipment costs.

$D = [d_i]$: Vector of length P, where d_i is the number of links plus β times the number of wavelength conversions used by lightpath i.

Minimize $D\,S + \alpha(\sum_{m,n} z_{mn} + \sum_{l,m} j_{lm})$ (10)

Subject to

28

$$\sum_{l,ws} y_{lmn}^{ws\ wd} > u_{mn}^{wd} \qquad (11)$$

$$\sum_{l,ws} y_{lmn}^{ws\ wd} < C_{max} u_{mn}^{wd} \qquad (12)$$

$$\sum_{n,wd} y_{lmn}^{ws\ wd} > v_{lm}^{ws} \qquad (13)$$

$$\sum_{n,wd} y_{lmn}^{ws\ wd} < C_{max} v_{lm}^{ws} \qquad (14)$$

$$\sum_{l,ws} y_{lmn}^{ws\ wd} + y_{mn}^{wd} - u_{mn}^{wd} = z_{mn} \qquad (15)$$

$$\sum_{m,wd} y_{lmn}^{ws\ wd} - v_{lm}^{ws} = j_{lm} \qquad (16)$$

Explanation of Equations: (10) provides the objective function which aims to minimize the costs associated with traffic grooming and wavelength conversion devices. (11) and (12) require the u variables to indicate if any multiplexing has occurred. (13) and (14) cause the v variables to indicate if any demultiplexing has occurred. (15) and (16) are just used to provide a smaller expression for the minimization function.

We believe that our mathematical formulation is very flexible and should be considered by network designers. This would give the option to route lightpaths through the optical network in a way that minimizes the cost of required traffic grooming and wavelength conversion devices. Careful traffic grooming allows conservation of wavelength resources so that more traffic can be added without the addition of new optical links. This allows one to keep an existing backbone all-optical network, and increase its capacity over that provided by wavelength routed networks that do not use traffic grooming or single-hop traffic grooming networks.

29

3.5 Lagrange Relaxation

Lagrangian relaxation is a widely used heuristic method for solving optimization problems. Relaxation methods are particularly good for generating bounds on the optimal solution to a given problem. In this section we provide a model that examines the use of Lagrangian relaxation as a tool for solving our GRWA problem.

Procedure used for solving the GRWA using Lagrange Relaxation

In our approach, sets of constraints are relaxed by adding them to the objective function with penalty coefficients, the Lagrangian multipliers. The objective is to dualize, possibly after a certain amount of remodeling, the constraints linking the components together in such a way that the original problem is transformed into disconnected and easier to solve sub-problems. The iterative procedure for solving a Lagrange problem is shown in Figure 3.1.

Objective Function

We can form a lagrangian relaxation for the GRWA cost problem by placing the complicating constraints from equation 17 in the objective function.

$$\text{Min } [DS + \alpha(\sum_{m,n} z_{mn} + \sum_{l,m} j_{lm}) - \lambda_{ij}(A^T X - \Lambda) - \rho_j(\sum_{1 \leq j \leq c} c_j col_j(X) . {}^*col_k(G^w) + C_{max})) \forall k, w, j \ ] \ (17)$$

Where λ and ρ are the multipliers.

Updating of the Lagrange multipliers

The quality of the bound generated depends greatly on the choice of multiplier values. Generally, some initial choice of multipliers is used to compute a first estimate of a lower bound. These multipliers are systematically modified in an iterative fashion to produce (hopefully) better bounds. The solution strategy we present to solve the GRWA problem comprises the following elements:

1. Begin with each multiplier at 0. Let the step size be some (problem dependent value) k.

2. Solve the minimization problem to get current solution x.

3. For every constraint violated by x, increase the corresponding muliplier by k.

4. For every constraint with positive slack relative to x, decrease the corresponding multilpier by k.

5. If m iterations have passed since the best relaxation value has decreased, cut k in half.

6. Go to 2.

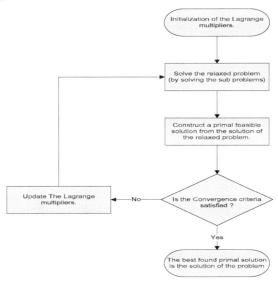

Figure 3.1: Procedure for solving the GRWA Lagrange problem

3.6 ILP Numerical Results

For figure 3.2 and figure 3.3, Table 3.1 presents the matrix of source-destination connection pairs that need to be established on the underlying optical network. In this section, we solve both the utilization and the cost problems for the given traffic table.

	1→4	1→6	2→4	3→5	4→3	5→6	6→4
OC-1	3	0	1	0	0	0	2
OC-12	0	2	0	1	2	1	0
OC-48	2	0	0	0	0	0	1

Table 3.1: The Traffic to Route on the Network

In this example, we assume that the maximum connection size is OC-*48* and that each link has two available wavelengths. The solution for the cost problem does not use any wavelength conversion (because the traffic grooming cost is much less than the wavelength conversion cost, and the connections can be routed without using wavelength conversion). Another observation is that traffic grooming is performed only on two of the nodes in the cost problem.

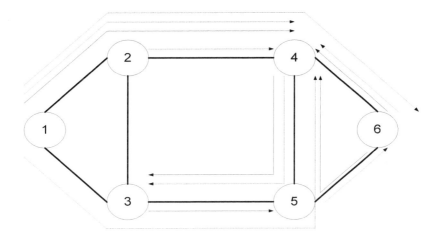

Figure 3.2: Routing and wavelength assignment for cost

On the other hand, the solution for the utilization problem does use wavelength conversion. Unlike the cost problem, in the utilization problem grooming and wavelength conversion are encouraged since we are trying to minimize the total number of wavelengths used in the network. We see that the utilization problem does favor grooming over using multiple wavelengths and the cost problem always chooses using multiple wavelengths (when available). Of course, the reason for this is that we have no associated cost for using multiple wavelengths instead of grooming, but grooming does have an associated cost.

To compare our example and solutions with those of others, we need to examine other methods of routing the connections. Since 5 connections have node *1* as their source, we could say that this example requires more than 2 wavelengths

33

unless there is at least end-to-end grooming. However, closer consideration shows that if we stipulate that we have no more than two available wavelengths, then there is contention for both links 2→4 and 3→5. The problem is that node *1* needs at least three connections to node *4* and one connection to node *6*, node *2* needs one connection to node *4*, and node *3* needs one connection to node *5* (that is, we need to route *5* connections over the two links which support only *4* total). We see that our example requires grooming in nodes other than end nodes, and grooming is not required on all of the nodes. In addition, wavelength conversion is not required on all of the nodes, and when the cost of wavelength conversion is higher than the grooming cost, grooming will be chosen over wavelength conversion. Another benefit is the amount of required grooming equipment. In the cost problem for this example we only need grooming equipment at two nodes.

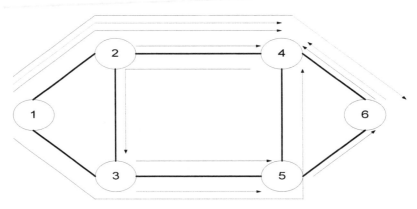

Figure 3.3: Routing and wavelength assignment for utilization

3.7 Summary

In this chapter, we introduced a mathematical formulation (ILP) in optical network with sparse traffic grooming and wavelength conversion resources under static traffic patterns. This formulation is very powerful and is very flexible for small networks under static traffic conditions. However, because the RWA problem is known to be NP-Complete, we know that the GRWA problem is NP-Complete. Thus, other approaches are presented in the next chapters for large-scale networks in terms of the number of nodes or the number of wavelengths.

CHAPTER 4

PROPOSED WAVELENGTH ASSIGNMENT HEURISTIC

4.1 Introduction

The wavelength assignment problem has been studies extensively. A summary of the research in this area can be found in [29]. A large number of wavelength assignment schemes have been proposed in the literature including random-fit, first-fit, most-used, least-used, least-loaded, min-product, max-sum, and relative capacity loss. These schemes can be classified into the following four categories [30]:

1. *Balance the load among all wavelengths*: These schemes usually perform poorly when compared to other wavelength assignment schemes (e.g., random-fit, least-used).

2. *Pack the wavelength usage*: These schemes are simple and perform well when the network state information is known precisely (e.g., first-fit, most-used).

3. *Spread the wavelength usage*: These schemes are also simple and perform as well as the schemes that pack the wavelength usage (e.g., least-loaded).

36

4. ***Global Assignment***: These schemes are more computationally extensive compared to the other schemes but they deliver the best performance (e.g., max-sum, relative capacity loss).

However, none of these wavelength assignment schemes account for the scarcity of the traffic grooming and wavelength conversion resources in backbone transport networks. In this chapter, we propose a simple GRWA heuristic that minimizes the use of traffic grooming and wavelength conversion resources as much as possible without hindering the blocking performance of the network. The rationale behind this is that the traffic grooming and wavelength conversion resources are very scarce and expensive resources in such networks and having a GRWA heuristic that conserves the usage of these resources is a critical requirement that can drastically conserve the usage of these resources without hindering the network blocking performance.

In this chapter we explain our objective function, and then we describe our heuristic to solve the GRWA problem in networks with sparse traffic grooming and wavelength conversion resources.

4.2 Objective Function

The GRWA problem with sparse traffic grooming resources presented in this chapter relies on the following assumptions:

1. The network is a general mesh topology with directed fiber connections. A pair of fiber links (i.e., one in each direction) is needed to connect a pair of nodes.

2. Network switches may or may not have support for traffic grooming.

3. Traffic grooming capability of each node is limited to the number of traffic grooming devices (resources) installed on that node.

4. Traffic grooming devices can perform wavelength conversion too but the cost a traffic grooming device is more than that of a wavelength conversion device since traffic grooming devices are capable of achieving more complex functionality (i.e., multiplexing and de-multiplexing connections). Thus, traffic grooming or wavelength conversion devices should be deployed in nodes based on whether traffic grooming is not needed or not.

5. Lightpaths do not contain loops. We use the K-shortest paths algorithm to enumerate the K shortest and loop-free paths between two nodes.

Our objective is to minimize the total cost of required wavelength conversion and traffic grooming hardware that needs to be installed in the network without hindering the blocking performance of the network. The total routing cost is represented as:

$$C = \sum_{i=1}^{M} D_i + \alpha \ G_i + \beta \ V_i \qquad (4.1)$$

Where:

- M : Number of lightpath requests.

- D_i : The number of hops for request i .

- G_i : The number of grooming devices used by request i .

- V_i : The number of wavelength conversions devices used by request i .

- α: The cost of a single traffic grooming device

- β : The cost of a single wavelength conversion device. It is assumed that α< β
  to reflect actual hardware cost.

- C: Total cost of routing all M lightpaths request though the optical network. This cost includes the cost of wavelengths used to carry the lightpath from its source to the destination node plus the cost of all wavelength conversion and traffic grooming devices used by the lightpath.

4.3 Proposed Most-Contiguous Heuristic Description

Our proposed heuristic strives to avoid wavelength conversion and wavelength bandwidth fragmentation by using paths with the most contiguous wavelength resources first. Figure 4.1 provides a flowchart of the proposed heuristic.

Most-Contiguous (MC) Heuristic

- **Definitions**

R: Number of requests.
GetFirstPathPointer: Function that returns a pointer to the first path maintained in the K-shortest patharray for the given request.
GetLastPathPointer: Function that returns a pointer to the last path maintained in the K-shortest path array for the given request.

AssignWavelengths: Function that handles wavelength assignment for the given path by saving the assigned wavelength for each link in a vector. This function returns *true* if the wavelength assignment succeeds, otherwise it returns *false*.

SavePath: Function that saves the path with its corresponding wavelength assignment.

GetSmallestPathCost. Function that returns the lowest cost path.

GetNumberOfHops: Function that returns the number of hops for the given path.

OR: Function that performs bitwise -or- operation of all the wavelength availability masks from start-hop to current-hop

MASK: Binary Vector of length equal to the number of wavelengths. Each bit in this vector reflects whether the individual wavelengths are used (1) or not (0).

AllUsed: Function that returns *true* if all the bits in MASK vector are used, otherwise it returns *false*.

SaveAssginWavelengths: Function that saves assigned wavelengths from start hop to current hop.

- **Pre-Processing**

  1: Generate uniform source-destination requests.

  2: Find K-Shortest Paths for every source-destination pair.

- **Main**

```
1: for each r from 1 to R
2:        firstPathPtr= GetFirstPathPointer (r)
3:        lastPathPtr=  GetLastPathPointer (r)
4:        for each path from firstPathPtr to lastPathPtr
5:                if( AssignWavelengths(path,
                                selectedWavelengths)==true )
6:                        SavePath (path, selectedWavelengths)
7:                end if
8:        end for
9:        SelectedPath= GetSmallestPathCost ()
11: end for
```

- **Wavelength Assignment**

  AssignWavelengths (pathPtr, selectedWavelengths)

  Start Procedure
  ```
  1: start =1
  2: current= 1
  3: N= GetNumberOfHops( pathPtr )
  4: While   ( current <= N )
  5:        while   ( true )
  6:                MASK = OR (start, current, pathPtr)
  7:                if   (AllUsed (MASK) )
  8:                       break
  9:                else
  10:               current=current+1
  11:        end if
  12:End while
  13:if (start == current)
  14:        return false
  15:else
  16:        SaveAssginWavelengths(MASK, start, current)
  17:        If ( current == N)
  18:                return true
  19:        else
  20:                start=current
  21:        end if
  22:end if
  23: end while
  24: return true
  ```
  End Procedure

  It should be noted here that the proposed algorithm conserves the traffic grooming and wavelength conversion resources as much as possible, however, when a tie occurs between multiple wavelength assignment options, any of the simple pack/spread wavelength assignment schemes presented above in the introduction can be used to break the tie. We suggest using the first-fit wavelength assignment scheme to break such ties because of the simplicity and good performance of this scheme. Also, notice that the algorithm proposed here does not guarantee that it will always find the wavelength assignment with the lowest possible number of traffic grooming and wavelength conversion devices. The algorithm strives to avoid wavelength bandwidth fragmentation in order to avoid increasing the network blocking performance. Also, the algorithm tries to keep the blocking performance as low as possible even at the expense of having more traffic grooming and/or wavelength conversion resources. A scheme that will always find the lowest number of traffic grooming and wavelength conversion resources can be computationally extensive and the scheme proposed here provides a good balance between simplicity and the efficiency of the found solutions.

To illustrate our most contiguous GRWA heuristic, let us assume that the following three lightpaths request need to be established on the network shown in Figure 4.2:

- *Lightpath 1*: *OC-3* From node *3* to node *4*

- *Lightpath 2*: *OC-3* from node *1* to node *5*.

- *Lightpath 3*: *OC-12* from node *2* to node *4*.

Assuming that the maximum capacity of a single wavelength is *OC-12*, our proposed algorithm will use a traffic grooming device on node *3* to multiplex lightpaths *1* and *2* on one wavelength while lightpath *3* will be carried over a separate wavelength since wavelength *1* does not have enough bandwidth to carry that lightpath as illustrated in Fig. 4.1a. If the first fit wavelength assignment heuristic is used, lightpaths *1* and *2* will be groomed on wavelength *1* using a grooming device on node *3* as before but lightpath 3 will use wavelength *1* on the WDM link from node *2* to node *3* and wavelength *2* on the WDM link from node *3* to node *4* using a wavelength conversion device on node *3* as illustrated in Figure 4.2

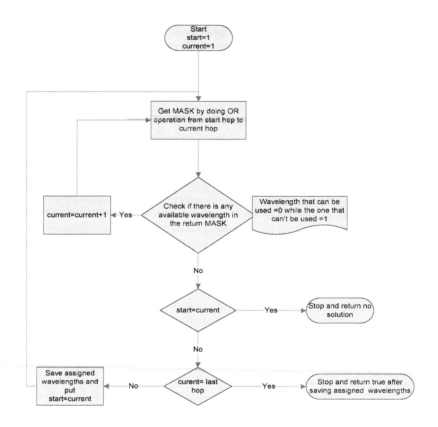

Figure 4.1: Most contiguous heuristic flowchart

44

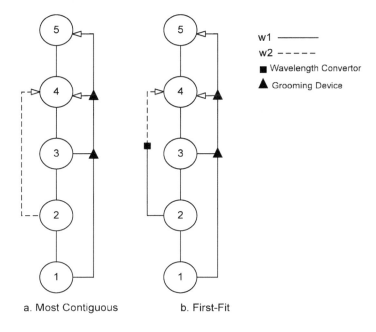

a. Most Contiguous   b. First-Fit

Figure 4.2: Explanation to the most contiguous

4.4 Summary

In this chapter, we proposed a simple and yet efficient heuristic for traffic grooming, routing, and wavelength assignment in optical mesh networks. A pseudocode and flowchart have been given for our proposed heuristic. This proposed approach efficiently uses the network resources by distributing the traffic more evenly among all network links, which has a major impact on lowering the blocking probability significantly. Performance results of our proposed heuristic will be demonstrated in Chapter 6.

CHAPTER 5

PROPOSED GENETIC APPROACH

5.1 Introduction

Previous traffic grooming studies decompose the GRWA problem into three sub-problems; namely: traffic grooming, wavelength, and route assignment problems. In this work, we employ a new approach that is based on the Genetic Algorithm (GA) to solve the GRWA problem. Our approach solves the traffic grooming, wavelength, and route assignment problems jointly without decomposing them into three separate problems. Our GA based approach is described in the following section.

5.2 Proposed Genetic Model Explanation

5.2.1   Chromosome Encoding

A chromosome is a vector of pointers to entries in the ***routing and wavelength assignment enumeration table***. The routing and wavelength assignment enumeration table enumerates all possible routing and wavelength assignment options for all given source-destination pairs. This table is generated by combining the *K*-Shortest routes for each source-destination pair with all the possible wavelength assignments for that route. Each unique wavelength assignment on a route is considered as a unique lightpath. Each gene on a chromosome represents one of those unique lightpaths for the given source-destination pair. The total length of the chromosome is equal to the number of lightpath requests presented to the networks.

To help understand our chromosome encoding technique, consider the example depicted in Figure 5.1 which represents a simple three node network. The figure shows an example of a two-gene chromosome that encodes two lightpaths. The first gene points to the 5th entry of the routing and wavelength assignment enumeration table while the second gene points to the 2nd entry of that table. Notice that the entries of the enumeration table have full routing and wavelength assignment information for the lightpath. For example, the enumeration table indicates that the 2nd entry uses wavelength 1 on the WDM link from node 1 to node 2 and wavelength 2 on the WDM link from node 2 to node 3. Also, the 5th entry of that table indicates that the lightpath that uses that entry will reserve wavelength 1 on the link from node 2 to node 3.

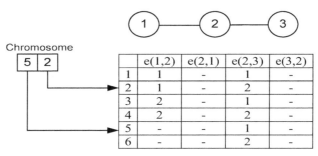

Figure 5.1: Chromosome encoding example using enumeration matrix

47

5.2.2 Initial Population

The first generation is formed from a combination of First-Fit, Most-Contiguous, and completely random chromosomes. In our model, the size of the initial population is *150* chromosomes (i.e., *50* chromosomes based on each of the three GRWA heuristics mentioned above).

5.2.3 Fitness Function

The fitness function of our GA model **F** includes a penalty component **P** as well as a cost component **C**. A high value γ is added to the value of the penalty component each time the selected route violates the number of traffic grooming resources, wavelength conversion resources, or the wavelength capacity constrains. In our formulation, we make the assumption that $\gamma >> (\alpha, \beta)$, where α and β represent the costs of single traffic grooming and wavelength conversion resources respectively.

The fitness function used in our model is defined as follows:

$$F = C + P$$

$$P = \gamma \sum_{i=1}^{M} \Re_i$$

$$\Re_i = \sum_{j=1}^{|L_i|} \sigma L_{ij}$$

$$\sigma = \begin{cases} 1 \text{ if Link } L_{ij} \text{ violates the capacity or the resources} \\ 0 \text{ otherwise} \end{cases}$$

Where:

- C : Same Objective function discussed in section 4.2

- M : Number of lightpath requests (chromosome length).

- L_i : The WDM links that the ith lightpath request traverses.

5.2.4   Crossover

In our model, crossover is performed between two parent chromosomes to produce two descendents using the two-point crossover technique. We chose the two-point crossover technique in our model in order to diversify the search within the large problem space.

5.2.5   Mutation

In our GA model, mutation is performed by walking through the genes that makeup the chromosome and modifying their value with a low probability (typically 0.1%). The resulting chromosomes need to be valid after mutation. If there is a chromosome that violates the routing constrains of a source-destination pair, we repair that chromosome by replacing the bad genes with valid ones in order to make a valid chromosome. The bad genes will be replaced by ones chosen from the list of valid genes based on a uniformly distributed selection process. This repair strategy guarantees that the gene will be selected from the range of enumerated lightpaths that belong to the given source-destination pair.

5.2.6   Selection

The chromosomes for crossover are chosen using the ***best selection method***. This selection method picks the best chromosome among the n chromosomes in a population in direct proportion to their absolute fitness. After crossover and mutation, new offsprings are reproduced then the best of those offsprings will be selected for the next generation. The offsprings with the worst fitness are discarded. The best selection method guarantees that the better chromosomes have a better chance to survive for the next generations. Figure 5.2 illustrates an example of our GA model when applied to Figure 5.1 In this figure, the chromosomes encode three lightpath requests as follows:

- **Lightpath 1**: From node *2* to node *3*.

- **Lightpath 2**: From node *1* to node *3*.

- **Lightpath 3**: From node *1* to node *3*.

In this example, after crossover, mutation, and applying the best selection method, we get a new chromosome for the same source-destination pairs, but without using any traffic grooming or wavelength conversion resources as can be seen from the routing and wavelength assignment enumeration matrix illustrated in Figure 5.1.

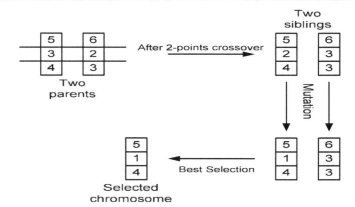

Figure 5.2: Illustration of the GA crossover, mutation and selection process used in our model

50

5.3 Summary

In this chapter, we introduced a GA model to solve the GRWA problem in optical networks with sparse traffic grooming and wavelength conversion resources. Our approach solves the traffic grooming, wavelength, and route assignment (GRWA) problems jointly without decomposing them into three separate problems. We presented the structure of our GA model that includes chromosome encoding, Initial population, fitness function, crossover operator, mutation operator, and selection method. Performance results of our proposed GA model will be demonstrated in Chapter 6.

CHAPTER 6

PERFORMANCE RESULTS

6.1 Introduction

The performance of our proposed Most-Contiguous and Genetic based heuristics has been compared with that of the first-fit GRWA approach in networks with sparse traffic grooming and wavelength conversion capabilities. We chose to compare our proposed heuristics with the first-fit heuristic because of the simplicity of this heuristic. Further, it was demonstrated in the literature that the first-fit heuristic produces low blocking probabilities [6].

The proposed heuristics were compared in terms of their blocking probability and total path cost in terms of used traffic grooming and wavelength conversion resources. In this chapter, we study the performance of our genetic algorithm, most-contiguous and first-fit for static traffic. In addition, we present the performance of most-contiguous and first-fit under dynamic traffic.

6.2 Analytical Results for Static Traffic Grooming

With the static traffic model, the generated lightpath requests are known ahead of time and are generated between all possible source-destination pairs with equal probabilities. This means that the source and destination nodes of all lightpath requests are chosen with uniform probabilities. The capacity of the generated lightpath requests also follows a uniform distribution between *1* and the maximum capacity of a single wavelength. Our simulation tool generates *n* lightpath requests to determine the blocking probability of the network and the total cost of the traffic grooming and wavelength conversion resources used by the offered lightpath requests.

We performed our performance evaluation study on the 16-node topology illustrated in Figure. 6.1. This figure reflects the structure of a reasonably complex mesh WDM transport network.

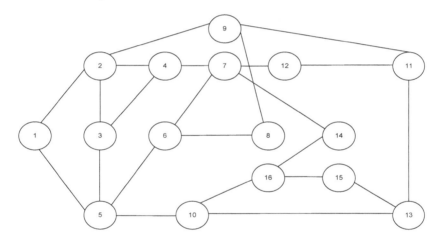

Figure 6.1: 16-node WDM mesh network

The performance of our proposed genetic based GRWA heuristic is evaluated for a population size of *150* chromosomes, crossover rate of *1*, and mutation rate of *%0.01* for a total of *150* epochs. Figures 6.2 though 6.4 plot the blocking probability versus the number of traffic grooming and wavelength conversion resources installed in the network for *70*, *100*, and *300* static lightpath requests, respectively. Those figures demonstrate that our genetic based GRWA approach achieves the best blocking probability performance under the different traffic loads compared to the most-contiguous and first-fit heuristics. The blocking performance of our most-contiguous heuristic is better than that of the first-fit heuristic. Particularly, Fig. 6.4 shows that our simple most contiguous heuristic can perform better than our genetic based GRWA approach under high traffic demands and low number of traffic grooming and wavelength conversion resources. Notice that Figures 6.2 through 6.4 indicate that the difference between the three heuristics is higher under low traffic demands and low number of traffic grooming and wavelength conversion resources. Figure 6.5 compares the total cost of traffic grooming and wavelength conversion resources used by the three GRWA heuristics in networks with various degrees of traffic grooming and wavelength conversion capabilities. The study shown in Figure 6.5 was conducted under the same blocking probability to make our comparison study fare and accurate. Again, we used the 16-node topology shown in Figure 6.1 to conduct this study. The maximum connection size is *OC-48* and each WDM link has four wavelengths. This study shows that the total cost of the traffic grooming and wavelength conversion resources used in our proposed most-contiguous and genetic based GRWA heuristics is much better than that of the first-fit heuristic. It should be emphasized here that our heuristics achieved lower costs without hindering the blocking performance of the network. Notice that the gap between our heuristics and the first-fit heuristic is higher in networks with sparse traffic grooming and wavelength conversion resources.

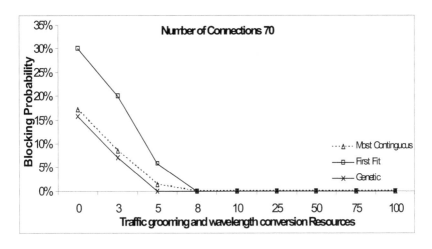

Figure 6.2: Blocking probability vs. number of traffic grooming and wavelength conversion resources using *70* lightpath requests

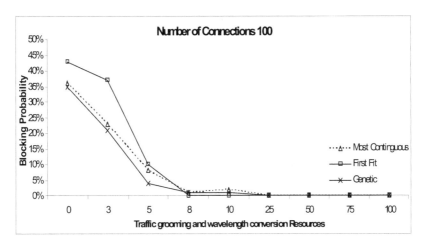

Figure 6.3: Blocking probability vs. number of traffic grooming and wavelength conversion resources using *100* lightpath requests

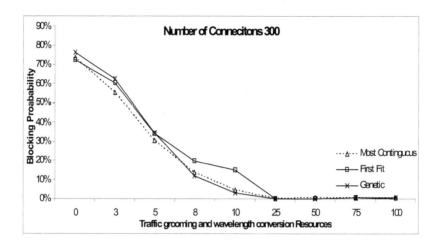

Figure 6.4: Blocking probability vs. number of traffic grooming and wavelength conversion resources using *300* lightpath requests

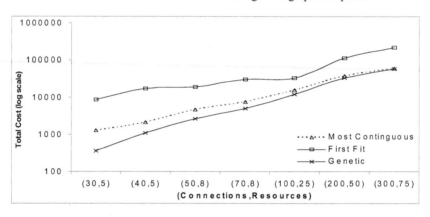

Figure 6.5: Total cost vs. number of connections, number of traffic grooming and wavelength conversion resources

6.3 Simulation Results for Dynamic Traffic Grooming

Extensive simulations have been carried out to investigate the performance of the proposed MC algorithm considering the same network topology depicted in Figure 6.1 for dynamic traffic. Each fiber link is assumed to carry 8 OC-48 wavelength channels. The flow dynamics of the network are modeled as follows:

1. The offered network load is given by:

$$L = \lambda H$$

    Where:

    L : Offered traffic load in Erlang.

    λ : Number of lightpath requests/ hour.

    H : Average call holding time in hours.

2. The connection holding time is exponentially distributed with mean $1/H$. We assume the holding time (H) to be 5 minutes.

3. Lightpath requests arrive at a node following an exponential distributed process with a mean $1/\lambda$. The destination node is uniformly chosen from all nodes except the source node of the lightpath.

4. The capacity of the lightpath requests follows a uniform distribution between OC-1 and the maximum capacity of a single wavelength.

Figure 6.6 compares the performance of the proposed most-contiguous heuristic with the first-fit heuristic under variable number of grooming and conversion resources. We observe that the most-contiguous heuristic significantly improves the blocking performance compared to the first-fit heuristic. In this study, we observed that when the first-fit heuristic is used, most of the traffic is distributed to the shortest route between each pair of nodes, resulting in congested links and the use of more grooming and conversion devices resources. On the other hand, our most-contiguous approach uses the network resources efficiently by distributing the traffic more evenly among all network links, which has a major impact on lowering the blocking probability significantly. This explains why first-fit in some cases outperforms the other approaches when there are large number of network resources.

Figure 6.6 indicates that increasing the number of grooming and conversion devices can significantly reduce the blocking probability for the most-contiguous as well as the first-fit heuristics especially when the network is heavily loaded. This could be explained with the fact that in the presence of more grooming and conversion devices, the algorithm is more likely to setup a lightpath for the source-destination pairs by utilizing the same resources to the extent possible.

In addition, Figure 6.6 illustrates that the blocking probability for a traffic load of *50* Erlangs is the same when the average number of traffic grooming and wavelength conversion resources is increases from *5* to *75*. This indicates that a network designer can reduce the network cost without affecting the network performance by carefully deploying a limited number of traffic grooming and wavelength conversion resources in the network.

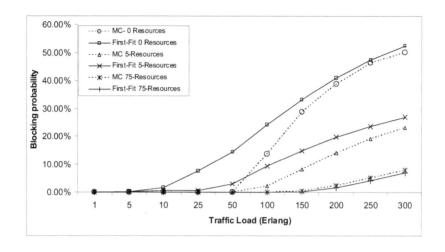

Figure 6.6: Comparison of the call blocking probability vs. traffic load of most
contiguous and first-fit using (0,5, and 75) traffic grooming and
wavelength conversion resources

Figure 6.7 shows the blocking probability with different number of traffic

grooming and wavelength conversion resources under a fixed heavy traffic load of

250 Erlangs. The purpose of this experiment is to study the performance implications

of using traffic grooming devices vs. wavelength conversion devices. We observe that

the performance of using traffic grooming devices only is much better than the

performance of using wavelength conversion devices (because traffic grooming

devices can also perform wavelength conversion but are more expensive). Also notice

that increasing the number of conversion devices under this heavy load has no major

impact on improving the blocking performance. This is due to the fact that the

resource bottleneck is the number of wavelengths on each fiber-link and not the number of wavelength converters at each node.

Figure 6.8 depicts the total cost of the paths selected by the most-contiguous and first-fit heuristics. As expected, we observe that the total cost of the most-contiguous approach is much better when compared to the first-fit heuristic. These results support our previous analysis under static traffic conditions [8].

Figure 6.9 shows that the difference between the average number of hops of the most-contiguous and first-fit heuristics is very small. This means that the most-contiguous heuristic can achieve a better cost than the first-fit heuristic without hindering the average number of hops.

Figure 6.11 studies the performance of having the traffic grooming and wavelength conversion devices on the edge nodes only (i.e., single-hop traffic grooming), compared to the case where the resources are distributed throughout the network. We use the 16-node network depicted in Figure 6.10, where we assume that nodes (1, 2, 5, 11, 13) are the edge nodes. Our results demonstrate that having the traffic grooming and wavelength conversion devices on the edge nodes only, can achieve very close blocking performance to the case of having them on every node. This Figure also indicates that the blocking performance does not always improve as the traffic grooming and wavelength conversion devices are placed throughout the optical network. This implies that a similar blocking performance can be achieved by deploying less traffic grooming and wavelength conversion devices on the edge nodes only.

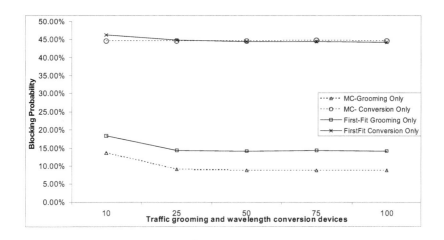

Figure 6.7: Comparison of the call blocking probability of resources that have traffic grooming capability only vs. resources that have wavelength conversion capability only

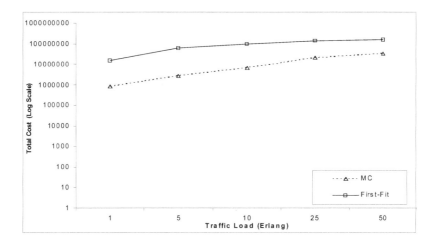

Figure 6.8: Comparison of total cost vs. traffic load of most contiguous and first-fit

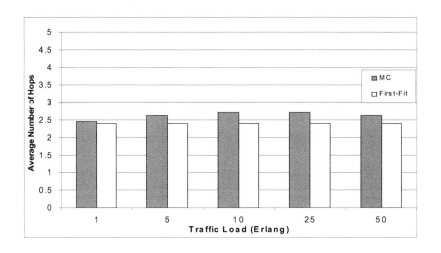

Figure 6.9: Comparison of average number of hops vs. traffic load of most contiguous and first-fit

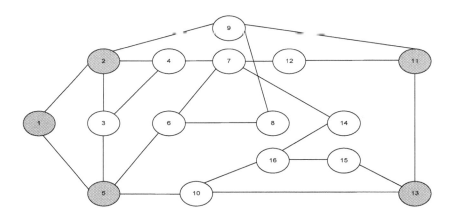

Figure 6.10: 16-node WDM mesh network where (1,2,5,11,13) are edge nodes

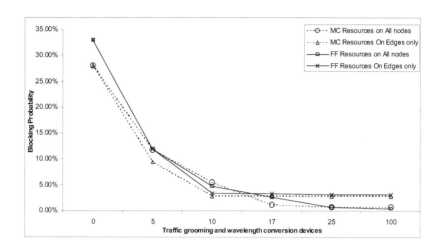

Figure 6.11: Comparison of the call blocking probability vs. traffic load of placing traffic grooming and wavelength conversion devices on edge nodes only and on all nodes

6.4 Summary

This chapter studied the performance of our GRWA heuristics under static and dynamic traffic. Our results show that our proposed heuristics reduce the total number of traffic grooming and wavelength conversion resources without hindering the blocking performance of the network. We compare the total cost and blocking performance of our proposed heuristic with that of the first fit heuristic and show that our proposed heuristics achieve better performance compared to the first fit heuristic approach. Our results also show that deploying traffic grooming and wavelength conversion resources on the edge of optical networks leads to lower cost networks with comparable blocking performance.

63

CHAPTER 7

CONCLUSION AND FUTURE WORK

7.1 Conclusions

In this work, we examined the problem of traffic grooming, routing, and wavelength assignment (GRWA) in WDM optical mesh networks with sparse traffic grooming and wavelength conversion resources under static and dynamic lightpath connection requests. First, the problem is formulated as an integer linear programming (ILP) problem. This ILP model is very powerful and is very flexible for small networks in terms of the number of nodes and the number of wavelengths. The GRWA problem is an NP-Complete since it is a generalization of the RWA problem which was proven to be NP-Complete. Thus, we propose two heuristic solutions to solve the GRWA problem in large-scale networks with sparse traffic grooming and wavelength conversion resources. Our first heuristic, strives to avoid wavelength conversion and bandwidth fragmentation by using paths with the most contiguous wavelength resources first. The second heuristic is an adaptation of the genetic algorithm to solve the GRWA problem in networks with sparse traffic grooming and wavelength conversion resources. The strength of the proposed heuristics stems from their simplicity, applicability to large-scale networks, and their efficiency compared to other heuristics proposed in the literature.

Our results demonstrate that our proposed heuristics reduce the total number of traffic grooming and wavelength conversion resources without hindering the blocking performance of the network. Moreover, our results also show that the blocking performance does not always improve as the traffic grooming and wavelength conversion devices are placed throughout the optical network. This implies that a network designer can reduce the network cost without affecting the network performance by carefully deploying a limited number of traffic grooming and wavelength conversion resources in the network.

7.2 Future Work

Areas of future work include the GRWA problem in optical mesh networks with protection requirements. Path protection approach requires finding a working path and a protection path that are link or node disjoint, so that the network is more survivable under various failures scenarios. Our proposed ILP formulations and heuristics can be extended to handle lightpath protection requirements. Furthermore, the performance of the proposed formulation and heuristics can be evaluated under such requirements.

Another attractive research problem is to design a multilayer sparse traffic grooming model. The main idea of this model is to have traffic grooming at the wavelength level then to group several wavelengths together as a band and switch the band using a single port whenever possible. To solve this problem, the ILP formulation, most-contiguous, and GA-based heuristics presented in this work need to be extended to handle optical bands.

BIBLIOGRAPHY

1. B. Mukherjee. "Optical communication networks". McGraw-Hill, 1997.

2. V. Alwayn. "Optical Network Design and implementation". Cisco-Press, 2004.

3. A.Gumaste and T.Antony. "DWDM Network Designs and Engineering Solutions". Cisco-Press 2002.

4. S. Song, "DWDM and the future integrated services networks," IEEE Canadian Review – Spring, no. 34, pp. 5-7, 2000.

5. K. Zhu and B. Mukherjee, "Traffic grooming in an optical WDM mesh network," IEEE Journal on Selected Areas in Communications, vol. 20, no. 1, pp.122-133, 2002.

6. J.M. Simmons E. L. Goldstein, and A. A. M. Saleh "Quantifying the Benefit of Wavelength Add-Drop in WDM Rings with Distance Independent and Dependent Traffic," IEEE/OSA Journal of Lightwave Tech., vol. 17, pp.48-57, Jan.1999.

7. R. Berry and E. Modiano. "Reducing electronic multiplexing costs in SONET/WDM rings with dynamically changing traffic," IEEE Journal on Selected Areas in Communications, vol. 18, no. 10, pp.1961-1971, 2000.

8. O. Gerstel, R. Ramaswami, and G. Sasaki. "Cost-effective traffic grooming in WDM rings," IEEE/ACM Transactions on Networking, vol. 8 no. 5, pp. 618-630, October 2000.

9. P-J. Wan, G. Calinescu, L. Liu, and O. Frieder. "Grooming of arbitrary traffic in SONET/WDM BLSRs," IEEE Journal on Selected Areas in Communications, vol. 18, no. 10, pp. 1995-2003, 2000.

10. R. Dutta and G. N. Rouskas, "On optimal traffic grooming in WDM rings," IEEE Journal on Selected Areas in Communications, vol. 20, no. 1, pp. 110-121, January 2002.

11. R. Dutta, S. Huang, and G. N. Rouskas, "Traffic grooming in path, star, and tree networks: Complexity, bounds, and algorithms," in Proc., ACM SIGMETRICS, pp. 298-299, June 2003.

12. J. Hu and B. Leida, "Traffic grooming, routing, and wavelength assignment in optical WDM mesh networks," in Proc., IEEE INFOCOM, 2004.

13. K. Zhu, H. Zang, and B. Mukherjee, "Design WDM mesh Networks with Sparse Grooming Capability," in Proc., IEEE Globecomm, Nov. 2002.

14. J. Bisschop. Introduction to all optical switching technologies. [Online]. Available: www.2cool4u.ch/wdm_dwdm/intro_allopticalswitching/ into_allopticalswitching.pdf. 2003, Jan.

15. Optical Internetworking Forum (OIF), http://www.oiforum.com.

16. R. Dutta, S. Huang, and G. N. Rouskas, "On optimal traffic grooming in elemental network topologies," in Proc of Opticomm 2003, October 2003.

17. A. Chiu and E. Modiano, "Traffic grooming algorithms for reducing electronic multiplexing costs in WDM ring networks," Journal of Lightwave Technology, vol. 18, no. 1, pp. 2-12, Jan 2000.

18. W. Yao and B. Ramamurthy, "Dynamic Traffic Grooming using Fixed-Alternate Routing in WDM Mesh Optical Networks," in the First Workshop on Traffic Grooming in WDM Networks, Co-located with BroadNets2004, October 2004.

19. A. Cox and J. Sanchez, "Cost Savings from Optimized Packing and L Grooming of Optical Circuits: Mesh versus Ring Comparisons, " Optical Network Magazine, vol. 2, no. 3, pp. 72-90, May/June 2001.

20. W. Yao, M. Li, and B. Ramamurthy, "Design of Sparse Grooming Networks for Transporting Dynamic Multi-Granularity Sub-Wavelength Traffic", In OFC/NFOEC 2005. Optical Fiber Communication Conference and Exposition

and the National Fiber Optical Engineering Conference, page OME68, Anaheim, CA, March 2005.

21. K. Zhu and B. Mukherjee, "A review of traffic grooming in WDM optical networks: Architectures and challenges, " SPIE Opt. Networks Mag, vol. 4, pp. 55–64, Mar./Apr. 2002

22. R. Barr and R. Patterson, "Grooming Telecommunication Networks," Optical Network Magazine, vol. 2, no. 3, pp. 20-23, May/June, 2001.

23. M. Sinclair, "Optical Mesh Topology Design using Node-Pair Encoding Genetic Programming" Genetic and Evolutionary Computation Conference (GECCO-99), Orlando, Florida, USA, pp.1192-1197, July 1999.

24. A. Smith, "Local Search Genetic Algorithm for Optimization of Highly Reliable Communications Networks, " pp.650-657 ICGA, 1997.

25. N. Banerjee and S. Sharan, "A Evolutionary Algorithm for Solving the Single Objective Static Routing and Wavelength Assignment Problem in WDM Networks, " in Proc. of ICISIP 2004, pp. 13-18, 2004.

26. H. Tzeng, J. Chen and N. Chen, "Traffic grooming in WDM networks using genetic algorithm," in Proc. IEEE SMC '99, vol. 1, 1999, pp. 1003-1006.

27. C. Lee and E. Park, "A genetic algorithm for traffic grooming in all-optical mesh networks," IEEE SMC 2002, volume 7, 6-9, Oct 2002.

28. Wikipedia the free encyclopedia. Available: http://en.wikipedia.org/wiki/Genetic_algorithm.

29. B. Mukherjee, "WDM optical communication networks: progress and challenges," IEEE Journal on Selected Areas in Communications, vol. 18, no.10, pp. 1810-1824, October, 2000.

30. J. Zhou and X. Yuan, "A study of dynamic routing and wavelength assignment with imprecise network state information," in Proc, International Conference on Parallel Processing Workshops, pp. 207-213, 2002.

31. G. Rouskas, H. Perros, "A Tutorial on Optical Networks," Networking Tutorial pp.155-194, 2002.

www.ingramcontent.com/pod-product-compliance
Lightning Source LLC
LaVergne TN
LVHW080103070326
832902LV00014B/2392